William Hamilton

An Account of the Late Eruption of Mount Vesuvius

In a Letter from the Right Honourable Sir William Hamilton, K. B. F. R. S. to

Sir Joseph Banks, Bart. P. R. S.

William Hamilton

An Account of the Late Eruption of Mount Vesuvius
*In a Letter from the Right Honourable Sir William Hamilton, K. B. F. R. S. to Sir
Joseph Banks, Bart. P. R. S.*

ISBN/EAN: 9783337196738

Printed in Europe, USA, Canada, Australia, Japan

Cover: Foto ©berggeist007 / pixelio.de

More available books at **www.hansebooks.com**

IV. *An Account of the late Eruption of Mount* Vesuvius. *In a Letter from the Right Honourable Sir* William Hamilton, *K. B. F. R. S. to Sir* Joseph Banks, *Bart. P. R. S.*

Read January 15, 1795.

SIR, Naples, August 25th, 1794.

EVERY day produces some new publication relative to the late tremendous eruption of mount Vesuvius, so that the various phænomena that attended it will be found on record in either one or other of these publications, and are not in that danger of being passed over and forgotten, as they were formerly, when the study of natural history was either totally neglected, or treated of in a manner very unworthy of the great Author of nature. I am sorry to say, that even so late as in the accounts of the earthquakes in Calabria in 1783, printed at Naples, nature is taxed with being malevolent, and bent upon destruction. In a printed account of another great eruption of Mount Vesuvius in 1631, by ANTONIO SANTORELLI, doctor of medicine, and professor of natural philosophy in the university of Naples, and at the head of the fourth chapter of his book, are these words: *Se questo incendio sia opera de' demonii? Whether this eruption be the work of devils?* The account of an eruption of Vesuvius in 1737, published at Naples by Doctor SERAO, is of a very different cast, and does great honour to his memory. All great eruptions of volcanoes must naturally produce nearly the same phænomena, and in SERAO's book almost all the phænomena we have been witness to during the late eruption of ,Vesuvius, are there admirably described, and well accounted for. The classical accounts of the eruption of Vesuvius, which destroyed the towns of Herculaneum and Pompeii, and many of the existing printed accounts of its great eruption in 1631 (although the latter are mixed with puerilities) might pass for an account of the late eruption by only changing the date, and omitting that circumstance of the retreat of the sea from the coast, which happened in both those great eruptions, and not in this; and I might content myself by referring to those accounts, and assuring you at the same time, that the late eruption, after those two, appears to have been the most violent recorded by history, and infinitely more alarming than either the eruption of 1767, or that of 1779, of both of which I had the honour of giving a particular account to the Royal Society. However, I think it my duty rather to hazard being guilty of repetition than to neglect the giving you every satisfaction in my power, relative to the late formidable operation of nature.

You know, Sir, that with the kind assistance of the Father ANTONIO PIAGGI, of the order of the *Scole Pie*, who has resided many years at Resina, on the very foot of Mount Vesuvius, and in the full view of it, I am in possession of an exact diary of that volcano, from the year 1779 to this day, and which is also accompanied with drawings. It is plain, from some remarks in that diary, previous to this eruption, that a great one was expected, and that we were apprehensive of the mischief that might probably attend the falling in of the crater, which had been much contracted within these two years past, by the great emission of scoriæ and ashes from time to time, and which had also increased the height of the volcano, and nearly filled up its crater. The frequent slight eruptions of lava for some years past have issued from near the summit, and ran in small channels in different directions down the flanks of the mountain, and from running in covered channels, had often an appearance as if they came immediately out of the sides of Vesuvius, but such lavas had not sufficient force to reach the cultivated parts at the foot of the mountain. In the year 1779, the whole quantity of the lava in fusion having been at once thrown up with violence out of the crater of Vesuvius, and a great part of it falling, and cooling on its cone, added much to the solidity of the walls of this huge natural chimney, if I may be allowed so to call it, and has not of late years allowed of a sufficient discharge of lava to calm that fermentation, which by the subterraneous noises heard at times, and by the explosions of scoriæ and ashes, was known to exist within the bowels of the volcano ; so that the eruptions of late years, before this last, have, as I have said, been simply from the lava having boiled over the crater, the sides being sufficiently strong to confine it, and oblige it to rise and overflow. The mountain had been remarkably quiet for seven months before its late eruption, nor did the usual smoke issue from its crater, but at times it emitted small clouds of smoke, that floated in the air in the shape of little trees. It was remarked by the Father ANTONIO DI PETRIZZI, a capuchin friar (who has printed an account of the late eruption) from his convent close to the unfortunate town of Torre del Greco, that for some days preceding this eruption a thick vapour was seen to surround the mountain, about a quarter of a mile beneath its crater, as it was remarked by him, and others at the same time, that both the sun and the moon had often an unusual reddish cast.

The water of the great fountain at Torre del Greco began to decrease some days before the eruption, so that the wheels of a corn-mill, worked by that water, moved very slowly; it was necessary in all the other wells of the town and its neighbourhood to lengthen the ropes daily, in order to reach at the water; and some of the wells became quite dry. Although most of the inhabitants were sensible of this phænomenon, not one of them seems to have suspected the true cause of it. It has been well attested, that eight days before the eruption, a man and two boys, being in a vineyard above Torre del Greco (and precisely on the spot where one of the new mouths opened, from whence the principal current of lava that destroyed the town issued), were much alarmed by a sudden puff of smoke that came out of the earth close to them, and was attended with a slight explosion.

Had this circumstance, with that of the subterraneous noises heard at Resina for two days before the eruption (with the additional one of the decrease of water in the wells, as above-mentioned) been communicated at the time, it would have required no great foresight to have been certain that an eruption of the volcano was near at hand, and that its force was directed particularly towards that part of the mountain.

On the 12th of June, in the morning, there was a violent fall of rain, and soon after the inhabitants of Resina, situated directly over the ancient town of Herculaneum, were sensible of a rumbling subterraneous noise, which was not heard at Naples.

From the month of January to the month of May last, the atmosphere was generally calm, and we had continued dry weather. In the month of May we had a little rain, but the weather was unusually sultry. For some days preceding the eruption, the Duke DELLA TORRE, a learned and ingenious nobleman of this country, and who has published two letters upon the subject of the late eruption, observed by his electrometers that the atmosphere was charged in excess with the electric fluid, and continued so for several days during the eruption : there are many other curious observations in the duke's account of the late eruption.

About 11 o'clock at night of the 12th of June, at Naples we were all sensible of a violent shock of an earthquake ; the undulatory motion was evidently from east to west, and appeared to me to have lasted near half a minute. The sky, which had been quite clear, was soon after covered with black clouds. The inhabitants of the towns and villages, which are very numerous at the foot of Vesuvius, felt this earthquake still more sensibly, and say, that the shock at first was from the bottom upwards, after which followed the undulation from east to west. This earthquake extended all over the Campagna Felice ; and their Sicilian Majesties were pleased to tell me, that the royal palace at Caserta, which is 15 miles from this city, and one of the most magnificent and solid buildings in Europe (the walls being 18 feet thick), was shook in such a manner as to cause great alarm, and that all the chamber bells rang. It was likewise much felt at Beneventum, about 30 miles from Naples ; and at Ariano in Puglia, which is at a much greater distance ; both these towns have been often afflicted with earthquakes.

On Sunday the 15th of June, soon after 10 o'clock at night, another shock of an earthquake was felt at Naples, but did not appear to be quite so violent as that of the 12th, nor did it last so long; at the same moment a fountain of bright fire, attended with a very black smoke and a loud report, was seen to issue, and rise to a great height, from about the middle of the cone of Vesuvius; soon after another of the same kind broke out at some little distance lower down; then, as I suppose by the blowing up of a covered channel full of red-hot lava, it had the appearance as if the lava had taken its course directly up the steep cone of the volcano. Fresh fountains succeeded one another hastily, and all in a direct line tending, for about a mile and a half down, towards the towns of Resina and Torre del Greco. I could count 15 of them, but I believe there were others obscured by the smoke. It seems probable, that all these fountains of fire, from their being in such an exact line, proceeded from one and the same long fissure down the flanks of the mountain, and that the lava and other volcanic matter forced its way out of the widest parts of the crack, and formed there the little mountains and craters that will be described in their proper place. It is impossible that any description can give an idea of this fiery scene, or of the horrid noises that attended this great operation of nature. It was a mixture of the loudest thunder, with incessant reports, like those from a numerous heavy artillery, accompanied by a continued hollow murmur, like that of the roaring of the ocean during a violent storm; and added to these was another blowing noise, like that of the going up of a large flight of sky-rockets, and which brought to my mind also that noise which is produced by the action of the enormous bellows on the furnace of the Carron iron foundery in Scotland, and which it perfectly resembled. The frequent falling of the huge stones and scoriæ, which were thrown up to an incredible height from some of the new mouths, and one of which having been since measured by the Abbé TATA (who has published an account of this eruption), was 10 feet high, and 35 in circumference, contributed undoubtedly to the concussion of the earth and air, which kept all the houses at Naples for several hours in a constant tremor, every door and window shaking and rattling incessantly, and the bells ringing. This was an awful moment! The sky, from a bright full moon and star-light, began to be obscured; the moon had presently the appearance of being in an eclipse, and soon after was totally lost in obscurity. The murmur of the prayers and lamentations of a numerous populace forming various processions, and parading in the streets, added likewise to the horror. As the lava did not appear to me to have yet a sufficient vent, and it was now evident that the earthquakes we had already felt had been occasioned by the air and fiery matter confined within the bowels of the mountain, and probably at no small depth (considering the extent of those earthquakes), I recommended to the company that was with me, who began to be much alarmed, rather to go and view the mountain at some greater distance, and in the open air, than to remain in the house, which was on the sea side, and in the part of Naples that is nearest and most exposed to Vesuvius. We accordingly went to Posilipo, and viewed the conflagration, now become still more considerable, from the sea side under that mountain; but whether from the eruption having increased, or from the loud reports of the volcanic explosions being repeated by the mountain

behind us, the noise was much louder, and more alarming
than that we had heard in our first position, at least a mile
nearer to Vesuvius. After some time, and which was about
two o'clock in the morning of the 16th, having observed that
the lavas ran in abundance freely, and with great velocity,
having made a considerable progress towards Resina, the town
which it first threatened, and that the fiery vapours which had
been confined had now free vent, through many parts of a
crack of more than a mile and a half in length, as was evident
from the quantity of inflamed matter and black smoke, which
continued to issue from the new mouths abovementioned without any interruption, I concluded that at Naples all danger from
earthquakes, which had been my greatest apprehension, was
now totally removed, and we returned to our former station at
S. Lucia at Naples.

All this time there was not the smallest appearance of fire
or smoke from the crater on the summit of Vesuvius; but the
black smoke and ashes issuing continually from so many new
mouths, or craters, formed an enormous and dense body of
clouds over the whole mountain, and which began to give
signs of being replete with the electric fluid, by exhibiting
flashes of that sort of zig-zag lightning, which in the volcanic language of this country is called *ferilli*, and which is
the constant attendant on the most violent eruptions. From
what I have read and seen, it appears to me, that the truest
judgment that can be formed of the degree of force of the
fermentation within the bowels of a volcano during its eruption, would be from observing the size, and the greater or
less elevation of those piles of smoky clouds, which rise out
of the craters, and form a gigantic mass over it, usually in the

form of a pine tree, and from the greater or less quantity of the *ferilli,* or volcanic electricity, with which those clouds appear to be charged.

During thirty years that I have resided at Naples, and in which space of time I have been witness to many eruptions of Vesuvius, of one sort or other, I never saw the gigantic cloud abovementioned replete with the electric fire, except in the two great eruptions of 1767, that of 1779, and during this more formidable one. The electric fire, in the year 1779, that played constantly within the enormous black cloud over the crater of Vesuvius, and seldom quitted it, was exactly similar to that which is produced, on a very small scale, by the conductor of an electrical machine communicating with an insulated plate of glass, thinly spread over with metallic filings, &c. when the electric matter continues to play over it in zigzag lines without quitting it. I was not sensible of any noise attending that operation in 1779; whereas the discharge of the electrical matter from the volcanic clouds during this eruption, and particularly the second and third days, caused explosions like those of the loudest thunder; and indeed the storms raised evidently by the sole power of the volcano, resembled in every respect all other thunder-storms; the lightning falling and destroying every thing in its course. The house of the Marquis of BERIO at S. Iorio, situated at the foot of Vesuvius, during one of these volcanic storms was struck with lightning, which having shattered many doors and windows, and damaged the furniture, left for some time a strong smell of sulphur in the rooms it passed through. Out of these gigantic and volcanic clouds, besides the lightning, both during this eruption and that of 1779, I have, with many others,

MDCCXCV. M

seen balls of fire issue, and some of a considerable mag-
nitude, which bursting in the air, produced nearly the same
effect as that from the air-balloons in fireworks, the electric
fire that came out having the appearance of the serpents with
which those firework balloons are often filled.   The day on
which Naples was in the greatest danger from the volcanic
clouds, two small balls of fire, joined together by a small link
like a chain-shot, fell close to my *casino*, at Posilipo; they
separated, and one fell in the vineyard above the house, and
the other in the sea, so close to it that I heard a splash in the
water; but, as I was writing, I lost the sight of this phænomenon, which was seen by some of the company with me, and
related to me as above.   The Abbé TATA, in his printed account of this eruption, mentions an enormous ball of this kind
which flew out of the crater of Vesuvius whilst he was standing on the edge of it, and which burst in the air at some distance from the mountain, soon after which he heard a noise
like the fall of a number of stones, or of a heavy shower of
hail.

During the eruption of the 15th at night, few of the inhabitants of Naples, from the dread of earthquakes, ventured to
go to their beds. The common people were either employed
in devout processions in the streets, or were sleeping on the
quays and open places; the nobility and gentry, having caused
their horses to be taken from their carriages, slept in them
in the squares and open places, or on the high roads just out
of the town.   For several days, whilst the volcanic storms of
thunder and lightning lasted, the inhabitants at the foot of the
volcano, both on the sea side and the Somma side, were often
sensible of a tremor in the earth, as well as of the concussions

in the air, but at Naples only the earthquakes of the 12th and
15th of June were distinctly and universally felt : this fair city
could not certainly have resisted long, had not those earthquakes been fortunately of a short duration. Throughout this
eruption, which continued in force about ten days, the fever
of the mountain, as has been remarked in former eruptions,
shewed itself to be in some measure periodical, and generally
was most violent at the break of day, at noon, and at mid-
night.

About four o'clock in the morning of the 16th, the crater of
Vesuvius began to shew signs of being open, by some black
smoke issuing out of it; and at daybreak another smoke, tinged
with red, issuing from an opening near the crater, but on the
other side of the mountain, and facing the town of Ottaiano,
shewed that a new mouth had opened there, and from which,
as we heard afterwards, a considerable stream of lava issued,
and ran with great velocity through a wood, which it burnt ;
and having run about three miles in a few hours, it stopped
before it had arrived at the vineyards and cultivated lands.
The crater, and all the conical part of Vesuvius, was soon involved in clouds and darkness, and so it remained for several
days ; but above these clouds, although of a great height, we
could often discern fresh columns of smoke from the crater,
rising furiously still higher, until the whole mass remained in
the usual form of a pine tree ; and in that gigantic mass of
heavy clouds the *ferilli*, or volcanic lightning, was frequently
visible, even in the day time. About five o'clock in the morning of the 16th we could plainly perceive, that the lava which
had first broke out from the several new mouths on the south

M 2

side of the mountain, had reached the sea, and was running
into it, having overwhelmed, burnt, and destroyed the greatest
part of Torre del Greco, the principal stream of lava having
taken its course through the very centre of the town. We observed from Naples, that when the lava was in the vineyards
in its way to the town, there issued often, and in different parts
of it, a bright pale flame, and very different from the deep red
of the lava ; this was occasioned by the burning of the trees
that supported the vines. Soon after the beginning of this
eruption, ashes fell thick at the foot of the mountain, all the
way from Portici to the Torre del Greco ; and what is remarkable, although there were not at that time any clouds in the
air, except those of smoke from the mountain, the ashes were
wet, and accompanied with large drops of water, which, as I
have been well assured, were to the taste very salt ; the road,
which is paved, was as wet as if there had been a heavy
shower of rain. Those ashes were black and coarse, like the
sand of the sea shore, whereas those that fell there, and at
Naples some days after, were of a light-grey colour, and as
fine as Spanish snuff, or powdered bark. They contained
many saline particles ; as I observed, when I went to the town
of Torre del Greco on the 17th of June, that those ashes that
lay on the ground, exposed to the burning sun, had a coat of
the whitest powder on their surface, which to the taste was
extremely salt and pungent. In the printed account of the
late eruption by EMANUEL SCOTTI, doctor of physic and professor of philosophy in the university of Naples, he supposes
(which appears to be highly probable) that the water which
accompanied the fall of the ashes at the beginning of the erup-

tion, was produced by the mixture of the inflammable and de-
phlogisticated air, according to experiments made by Doctor
PRIESTLEY and Monsieur LAVOISIER.

By the time that the lava had reached the sea, between five
and six o'clock in the morning of the 16th, Vesuvius was so completely involved in darkness, that we could no more discern the
violent operation of nature that was going on there, and so it
remained for several days ; but the dreadful noise we heard at
times, and the red tinge on the clouds over the top of the
mountain, were evident signs of the activity of the fire underneath. The lava ran but slowly at Torre del Greco after it
had reached the sea ; and on the 17th of June in the morning, when I went in my boat to visit that unfortunate town,
its course was stopped, excepting that at times a little rivulet
of liquid fire issued from under the smoking scoriæ into the
sea, and caused a hissing noise, and a white vapour smoke ; at
other times, a quantity of large scoriæ were pushed off the
surface of the body of the lava into the sea, discovering that
it was red hot under that surface ; and even to this day the
centre of the thickest part of the lava that covers the town re-
tains its red heat.   The breadth of the lava that ran into the
sea, and has formed a new promontory there, after having destroyed the greatest part of the town of Torre del Greco,
having been exactly measured by the Duke DELLA TORRE, is
of English feet 1204. Its height above the sea is 12 feet, and
as many feet under water ; so that its whole height is 24 feet ;
it extends into the sea 626 feet. I observed that the sea water
was boiling as in a cauldron, where it washed the foot of this
new formed promontory ; and although I was at least an hundred yards from it, observing that the sea smoked near my

boat, I put my hand into the water, which was literally
scalded ; and by this time my boatmen observed that the pitch
from the bottom of the boat was melting fast, and floating
on the surface of the sea, and that the boat began to leak ; we
therefore retired hastily from this spot, and landed at some distance from the hot lava. The town of Torre del Greco contained about 18000 inhabitants, all of which (except about 15,
who from either age or infirmity could not be moved, and were
overwhelmed by the lava in their houses) escaped either to
Castel-a-mare, which was the ancient Stabiæ, or to Naples;
but the rapid progress of the lava was such, after it had al-
tered its course from Resina, which town it first threatened,
and had joined a fresh lava that issued from one of the new
mouths in a vineyard, about a mile from the town, that it ran
like a torrent over the town of Torre del Greco, allowing the
unfortunate inhabitants scarcely time to save their lives ; their
goods and effects were totally abandoned, and indeed several of
the inhabitants, whose houses had been surrounded with lava
whilst they remained in them, escaped from them and saved
their lives the following day, by coming out of the tops of their
houses, and walking over the scoriæ on the surface of the red-
hot lava. Five or six old nuns were taken out of a convent in
this manner, on the 16th of June, and carried over the hot
lava, as I was informed by the friar who assisted them; and
who told me that their stupidity was such, as not to have been
the least alarmed, or sensible of their danger : he found one
of upwards of 90 years of age actually warming herself at a
point of red-hot lava, which touched the window of her cell,
and which she said was very comfortable ; and though now
apprized of their danger, they were still very unwilling to leave

the convent, in which they had been shut up almost from their
infancy, their ideas being as limited as the space they inhabited. Having desired them to pack up whatever they had
that was most valuable, they all loaded themselves with bis-
cuits and sweetmeats, and it was but by accident that the friar
discovered that they had left a sum of money behind them,
which he recovered for them ; and these nuns are now in a
convent at Naples.

At the time I landed at Torre del Greco on the 17th, I
found some few of its inhabitants returned, and endeavouring
to recover their effects from such houses as had not been
thrown down, or were not totally buried under the lava; but
alas ! what was their cruel disappointment when they found
that their houses had been already broke open, and completely gutted of every thing that was valuable ; and I saw
a scuffle at the door of one house, between the proprietors,
and the robbers who had taken possession of it. The lava
had passed over the centre and best part of the town ; no part
of the cathedral remained above it, except the upper part of a
square brick tower, in which are the bells ; and it is a curious
circumstance that those bells, although they are neither cracked
or melted, are deprived of their tone as much as if they had
been cracked, I suppose by the action of the acid and vitriolic
vapours of the lava. Some of the inhabitants of Torre del
Greco told me, that when the lava first entered the sea, it
threw up the water to a prodigious height ; and particularly
when two points of lava met and inclosed a pool of water,
that then that water was thrown up with great violence, and
a loud report : they likewise told me, that at this time, as well
as the day after, a great many boiled fish were seen floating

on the surface of the sea; and I have since been assured by many of the fishermen of Portici, Torre del Greco, and Torre dell' Annunziata (all of which towns are situated at the foot of Vesuvius), that they could not for many days during the eruption catch a fish within two miles of that coast, which they had evidently deserted.

When this lava is cooled sufficiently, which may not be until some months hence, I shall be curious to examine whether the centre, or solid and compact parts, of the lava that ran into the sea has taken, as it probably may, the prismatical form of basalt columns, like many other ancient lavas disgorged into the water. The exterior of this lava at present, like all others, offers to the eye nothing but a confused heap of loose scoriæ. The lava over the cathedral, and in other parts of the town, is upwards of 40 feet in thickness; the general height of the lava during its whole course is about 12 feet, and in some parts not less than a mile in breadth. I walked in the few remaining streets of the town, and I went on the top of one of the highest houses that was still standing, although surrounded by the lava; I saw from thence distinctly the whole course of the lava, that covered the best part of the town; the tops of the houses were just visible here and there in some parts, and the timbers within still burning caused a bright flame to issue out of the surface; in other parts, the sulphur and salts exhaled in a white smoke from the lava, forming a white or yellow crust on the scoriæ round the spots where it issued with the most force. Often I heard little explosions, and saw that they blew up, like little mines, fragments of the scoriæ and ashes into the air; I suppose them to have been occasioned either by rarefied air in confined cellars, or perhaps by small portions of gunpowder taking fire, as few in this country are without a gun and some little portion of gunpowder in their houses. As the church feasts are here usually attended with fireworks and crackers, a firework-maker of this town had a very great quantity of fireworks ready made for an approaching feast, and some gunpowder, all of which had been shut up in his house by the lava, a part of which had even entered one of the rooms ; yet he actually saved all his fireworks and gunpowder some days after, by carrying them safely over the hot lava. I should not have been so much at my ease had I known of this gunpowder, and of several other barrels that were at the same time in the cellar of another house, inclosed by the lava, and which were afterwards brought off on women's heads, little thinking of their danger, over the scoriæ of the lava, that was red-hot underneath. The heat in the streets of the town, at this time, was so great as to raise the quicksilver of my thermometer to very near 100 degrees, and close to the hot lava it rose much higher ; but what drove me from this melancholy spot was, that one of the robbers with a great pig on his shoulders, pursued by the proprietor with a long gun pointed at him, kept dodging round me to save himself ; I bid him throw down the pig and run, which he did ; and the proprietor, satisfied with having recovered his loss, acquainted me with my danger, by telling me that there were now thieves in every house that was left standing. I thought it therefore high time to retire, both for my own safety, and that I might endeavour to procure from Naples some protection for the doubly unfortunate sufferers of this unhappy town. Accordingly I returned to Naples in my boat, and immediately acquainted this government with what I had just seen myself ;

in consequence of which a body of soldiers was sent directly
to their relief by sea, the road by land having been cut off by
the lava. I remarked in my way home, that there was a much
greater quantity of the petroleum floating on the surface of the
sea, and diffusing a very strong and offensive smell, than was
usual ; for at all times in calms, patches of this bituminous oil,
called here petroleum, are to be seen floating on the surface of
the sea between Portici and Naples, and particularly opposite
a village called Pietra Bianca. The minute ashes continued
falling all this day at Naples ; the mountain, totally obscured
by them, continued to alarm us with repeated loud explosions ;
the streets of this city were this day and the next constantly
filled with religious and penitential processions, composed of
all classes, and nothing was heard in the midst of darkness but
the thunder of the mountain, and *ora pro nobis*. The sea
wind increasing at times, delivered us from these ashes, which
it scattered over different parts of the Campagna Felice.

On Wednesday the 18th, the wind having for a very short
space of time cleared away the thick cloud from the top of
Vesuvius, we discovered that a great part of its crater, particularly on the west side opposite Naples, had fallen in, which it
probably did about four o'clock in the morning of this day, as
a violent shock of an earthquake was felt at that moment at
Resina, and other parts situated at the foot of the volcano.
The clouds of smoke, mixed with the ashes which, as I have
before remarked, were as fine as Spanish snuff (so much so
that the impression of a seal with my coat of arms would remain distinctly marked upon them), were of such a density as
to appear to have the greatest difficulty in forcing their pas-
sage out of the now widely extended mouth of Vesuvius, which

certainly, since the top fell in, cannot be much short of two miles in circumference. One cloud heaped on another, and succeeding one another incessantly, formed in a few hours such a gigantic and elevated column of the darkest hue over the mountain, as seemed to threaten Naples with immediate destruction, having at one time been bent over the city, and appearing to be much too massive and ponderous to remain long suspended in the air; it was besides replete with the *ferilli,* or volcanic lightning, which was stronger than common lightning, just as PLINY the younger describes it in one of his letters to TACITUS, when he says *fulgoribus illæ et similes et majores erant.*

Vesuvius was at this time completely covered, as were all the old black lavas, with a thick coat of these fine light-grey ashes already fallen, which gave it a cold and horrid appearance; and in comparison of the abovementioned enormous mass of clouds, which certainly, however it may contradict our idea of the extension of our atmosphere, rose many miles above the mountain, it appeared like a mole-hill; although, as you know, Sir, the perpendicular height of Vesuvius from the level of the sea, is more than three thousand six hundred feet. The Abbé BRACCINI, as appears in his printed account of the eruption of Mount Vesuvius in 1631, measured with a quadrant the elevation of a mass of clouds of the same nature, that was formed over Vesuvius during that great eruption, and found it to exceed thirty miles in height. Doctor SCOTTI, in his printed account of this eruption, says that the height of this threatening cloud of smoke and ashes, measured (but he does not say how) from Naples, was found to be of an elevation of thirty degrees. All I can say is, that to my eye

the distance from the crater of Vesuvius to the most elevated
part of the cloud, appeared to me nearly the same as that of
the island of Caprea from Naples, and which is about 25 miles;
but I am well aware of the inaccuracy of such a sort of mea-
surement.   At the time of its greatest elevation, I engaged
Signor GATTA, successor to the late ingenious Mr. FABRIS,
to make an exact drawing of it, which he did with great success ; and a copy of that drawing on a small scale is inclosed
(Tab. VII.), and will, I hope, give you a very good idea of
what I have been describing.

   I must own, that at that moment I did apprehend Naples
to be in some danger of being buried under the ashes of
the volcano, just as the towns of Herculaneum and Pompeii
were in the year 79.   The ashes that fell then at Pompeii
were of the same fine quality as those from this eruption ;
having often observed, when present at the excavations of
that ancient city, that the ashes, which I suppose to have
been mixed with water at the same time, had taken the exact impression or mould of whatever they had inclosed ; so
that the compartments of the wood work of the windows
and doors of the houses remained impressed on this volcanic
tufo, although the wood itself had long decayed, and not an
atom of it was to be seen, except when the wood had been
burnt, and then you found the charcoal.   Having once been
present at the discovery of a skeleton in the great street of
Pompeii, of a person who had been shut up by the ashes during
the eruption of 79, I engaged the men that were digging to
take off the piece of hardened tufo, that covered the head, with
great care, and, as in a mould just taken off in plaster of Paris,
we found the impression of the eyes, that were shut, of the

nose, mouth, and of every feature perfectly distinct. A similar specimen of a mould of this kind, brought from Pompeii, is now in his Sicilian Majesty's museum at Portici ; it had been formed over the breast of a young woman that had been shut up in the volcanic matter ; every fold of a thin drapery that covered her breast is exactly represented in this mould : and in the volcanic tufo that filled the ancient theatre of Herculaneum, the exact mould or impression of the face of a marble bust is still to be seen, the bust or statue having been long since removed. Having observed these fine ashes issuing in such abundance from Vesuvius, and having the appearance of being damp or wet, as you may perceive by the drawing (Tab. VII.) that they do not take such beautiful forms and volutes as a fine dry smoke usually does, but appear in harsh and stiff little curls, you will not wonder then, that the fate of Herculaneum and Pompeii should have come again strongly into my mind ; but fortunately the wind sprung up fresh from the sea, and the threatening cloud bent gradually from us over the mountain of Somma, and involved all that part of the Campagna in obscurity and danger.

To avoid prolixity and repetition, I need only say, that the storms of thunder and lightning, attended at times with heavy falls of rain and ashes, causing the most destructive torrents of water and glutinous mud, mixed with huge stones, and trees torn up by the roots, continued more or less to afflict the inhabitants on both sides of the volcano until the 7th of July, when the last torrent destroyed many hundred acres of cultivated land, between the towns of Torre del Greco and Torre dell' Annunziata. Some of these torrents, as I have been credibly assured by eye witnesses, both on the

sea side and the Somma side of the mountain, came down
with a horrid rushing noise; and some of them, after having forced their way through the narrow gullies of the mountain, rose to the height of more than 20 feet, and were near
half a mile in extent. The mud of which the torrents were
composed, being a kind of natural mortar, has completely
cased up, and ruined for the present, some thousand acres
of rich vineyards; for it soon becomes so hard, that nothing
less than a pick-axe can break it up; I say for the present,
as I imagine that hereafter the soil may be greatly improved
by the quantity of saline particles that the ashes from this
eruption evidently contain. A gentleman of the British fac-
tory at Naples, having filled a plate with the ashes that had
fallen on his balcony during the eruption, and sowed some
pease in them, assured me that they came up the third day,
and that they continue to grow much faster than is usual in
the best common garden soil.

My curiosity, or rather my wish to gratify that of our re-
spectable Society, induced me to go upon Mount Vesuvius, as
soon as I thought I might do it with any degree of prudence,
which was not until the 30th of June, and then it was attended
with some risk, as will appear in the course of this narrative.
The crater of Vesuvius, except at short intervals, had been
continually obscured by the volcanic clouds ever since the
16th, and was so this day, with frequent flashes of lightning
playing in those clouds, and attended as usual with a noise
like thunder; and the fine ashes were still falling on Vesuvius,
but still more on the mountain of Somma. I went up the
usual way by Resina, attended by my old Cicerone of the
mountain, BARTOLOMEO PUMO, with whom I have been

sixty-eight times on the highest point of Vesuvius. I observed
in my way through the village of Resina that many of the
stones of the pavement had been loosened, and were deranged
by the earthquakes, particularly by that of the 18th, which
attended the falling in of the crater of the volcano, and which,
as they told me there, had been so violent as to throw many
people down, and obliged all the inhabitants of Resina to quit
their houses hastily, and to which they did not dare return for
two days. The leaves of all the vines were burnt by the ashes
that had fallen on them, and many of the vines themselves
were buried under the ashes, and great branches of the trees
that supported them had been torn off by their weight. In
short, nothing but ruin and desolation was to be seen. The
ashes at the foot of the mountain were about 10 or 12 inches
thick on the surface of the earth, but in proportion as we ascended their thickness increased to several feet, I dare say
not less than 9 or 10 in some parts ; so that the surface of the
old rugged lavas, that before was almost impracticable, was
now become a perfect plain, over which we walked with the
greatest ease. The ashes were of a light-grey colour, and exceedingly fine, so that by the footsteps being marked on them
as on snow, we learnt that three small parties had been up before us. We saw likewise the track of a fox, that appeared to
have been quite bewildered, to judge from the many turns he
had made. Even the traces of lizards and other little animals,
and of insects, were visible on these fine ashes. We ascended to
the spot from whence the lava of the 15th first issued, and we
followed the course of it, which was still very hot (although
covered with such a thick coat of ashes), quite down to the
sea at Torre del Greco, which is more than five miles. A pair

of boots, to which I had for the purpose added a new and thick sole, were burnt through on this expedition. It was not possible to get up to the great crater of Vesuvius, nor had any one yet attempted it. The horrid chasms that exist from the spot where the late eruption first took place, in a straight line for near two miles towards the sea, cannot be imagined. They formed vallies more than two hundred feet deep, and from half to a mile wide ; and where the fountains of fiery matter existed during the eruption, are little mountains with deep craters. Ten thousand men, in as many years, could not, surely, make such an alteration on the face of Vesuvius, as has been made by nature in the short space of five hours. Except the exhalations of sulphureous and vitriolic vapours, which broke out from different spots of the line abovementioned, and tinged the surface of the ashes and scoriæ in those parts with either a deep or pale yellow with a reddish ochre colour, or a bright white, and in some parts with a deep green and azure blue (so that the whole together had the effect of an iris), all around us had the appearance of a sandy desert. We went on the top of seven of the most considerable of the new-formed mountains, and looked into their craters, which on some of them appeared to be little short of half a mile in circumference ; and although the exterior perpendicular height of any of them did not exceed two hundred feet, the depth of their inverted cone within was three times as great. It would not have been possible for us to have breathed on these new mountains near their craters, if we had not taken the precaution of tying a doubled handkerchief over our mouths and nostrils; and even with that precaution we could not resist long, the fumes of the vitriolic acid were so exceedingly penetrating, and of such a suffocating quality. We found in one a double crater, like two funnels joined together; and in all there was some little smoke and depositions of salts and sulphurs, of the various colours above mentioned, just as is commonly seen adhering to the inner walls of the principal crater of Vesuvius.

Two or three days after we had been here, one of the new mouths into which we had looked, suddenly made a great explosion of stones, smoke, and ashes, which would certainly have proved fatal to any one who might unfortunately have been there at the time of the explosion. We read of a like accident having proved fatal to more than twenty people, who had the curiosity to look into the crater of the Monte Nuovo, near Pozzuoli, a few days after its formation, in the year 1538. The 15th of August, I saw a sudden explosion of smoke and ashes, thrown to an extreme height out of the great crater of Vesuvius, that must have destroyed any one within half a mile of it; and yet on the 19th of July a party not only had visited that crater, but had descended 170 feet within it. Whilst we were on the mountain, two whirlwinds, exactly like those that form water-spouts at sea, made their appearance; and one of them that was very near us made a strange rushing noise, and having taken up a great quantity of the fine ashes, formed them into an elevated spiral column, which, with a whirling motion and great rapidity, was carried towards the mountain of Somma, where it broke and was dispersed. As there were evident signs of an abundance of electricity in the air at this time, I have no doubt of this having been also an electrical operation. One of my servants, employed in collecting of sulphur, or sal ammoniac, which crystallizes near the *fumaroli,* as they are called here (and which are the spots from whence

the hot vapour issues out of the fresh lavas), found to his great
surprise an exceeding cold wind issue from a fissure very near
the hot *fumaroli* abovementioned upon his leg; I put my hand
to the spot, and found the same; but it did not surprise me, as
before on Mount Vesuvius, on the mountain of Somma, on
Mount Etna, and in the island of Ischia, I had met with, on
particular spots, the like currents of extreme cold air issuing
from beneath the ancient lavas, and which, being constant to
those spots, are known by the name of *ventoroli*. In a vineyard not in the same line with the new-formed mountains just
described, but in a right line from them, at the distance of
little more than a mile from Torre del Greco, are three or
four more of these new-formed mountains with craters, out of
which the lava flowed, and by uniting with the streams that
came from the higher mouths, and adding to their heat and
fluidity, enabled the whole current to make so rapid a progress over the unfortunate town, as scarcely to allow its inhabitants sufficient time to escape with their lives. The rich
vineyards belonging to the Torre del Greco, and which produced the good wine called *Lacrima Christi*, that have been
buried, and are totally destroyed by this lava, consisted, as I
have been informed, of more than three thousand acres; but
the destruction of the vineyards by the torrents of mud and
water at the foot of the mountain of Somma, is much more
extensive.

I visited that part of the country also a few days after I had
been on Vesuvius, not being willing to relate to you any one
circumstance of the late formidable eruption but what I had
reason to believe was founded on truth. The first signs of a
torrent that I met with, was near the village of the Madonna

dell' Arco, and I passed several others between that and the town of Ottaiano ; the one near Trochia, and two near the town of Somma, were the most considerable, and not less than a quarter of a mile in breadth ; and as several eye witnesses assured me on the spot, were, when they poured down from the mountain of Somma, from 20 to 30 feet high ; it was a liquid glutinous mud, composed of scoriæ, ashes, stones (some of which of an enormous size) mixed with trees that had been torn up by the roots. Such torrents, as you may well ima- gine, were irresistible, and carried all before them ; houses, walls, trees, and, as they told me, not less than four thousand sheep and other cattle, had been swept off by the several tor- rents on that side of the mountain. At Somma they likewise told me that a team of eight oxen, that were drawing a large timber tree, had been carried off from thence, and never were more heard of.

The appearance of these torrents, when I saw them, was like that of all other torrents in mountainous countries, except that what had been mud was become a perfect cement, on which nothing less than a pick-axe could make any impression. The vineyards and cultivated lands were here much more ruined ; and the limbs of the trees much more torn by the weight of the ashes, than those which I have already de- scribed on the sea side of the volcano.

The Abbé TATA, in his printed account of this eruption, has given a good idea of the abundance, the great weight, and glutinous quality of these ashes, when he says that having taken a branch from a fig-tree still standing near the town of Somma, on which were only six leaves, and two little unripe figs, and having weighed it with the ashes attached to it, he found it to be 31 ounces ; when having washed off the volcanic matter, it scarcely weighed 3 ounces.

I saw several houses on the road, in my way to the town of Somma, with their roofs beaten in by the weight of the ashes. In the town of Somma, I found four churches and about seventy houses without roofs, and full of ashes. The great damage on this side of the mountain, by the fall of the ashes and the torrents, happened on the 18th, 19th, and 20th of June, and on the 12th of July. I heard but of three lives that had been lost at Somma by the fall of a house. The 19th, the ashes fell so thick at Somma (as they told me there), that unless a person kept in motion, he was soon fixed to the ground by them. This fall of ashes was accompanied also with loud reports, and frequent flashes of the volcanic lightning, so that, surrounded by so many horrors, it was impossible for the inhabitants to remain in the town, and they all fled ; the darkness was such, although it was mid-day, that even with the help of torches it was scarcely possible to keep in the high road ; in short, what they described to me was exactly what PLINY the younger and his mother had experienced at Misenum during the eruption of Vesuvius in the reign of TITUS, according to his second letter to TACITUS on that subject. I found that the majority of people here were convinced that the torrents of mud and water, that had done them so much mischief, came out of the crater of Vesuvius, and that it was sea-water ; but there cannot be any doubt of those floods having been occasioned by the sudden dissolution of watery clouds mixed with ashes, the air perhaps having been too much rarefied to support them ; and when such clouds broke, and fell heavily on Vesuvius, the water not being able to penetrate

as usual into the pores of the earth, which were then filled up
with the fine ashes of a bituminous and oily quality, nor having free access to the channels which usually carried it off, accumulated in pools, and mixing with more ashes, rose to a
great height, and at length forced its way through new channels, and came down in torrents over countries where it was
least expected, and spread itself over the fertile lands at the
foot of the mountain. From what I have seen lately, I begin to
doubt very much if the water, by which so much damage was
done, and so many lives were lost during the terrible eruption
of Vesuvius in 1631, did really, as was generally supposed,
come out of the crater of the volcano : sentiments were divided
then, as they are now, on that subject ; and since in all great
eruptions the crater of the volcano must be obscured by the
clouds of ashes, as it probably was then, and certainly was
during the violence of the late eruption, therefore it must be
very difficult to ascertain exactly from whence that water came.
The more extraordinary a circumstance is, the more it appears
to be the common desire that it should be credited ; from this
principle, one of his Sicilian Majesty's gardeners of Portici
went up to the crater of Vesuvius as soon as it was practicable,
and came down in a great fright, declaring that he had seen
it full of boiling water. The Chevalier MACEDONIO, intendant
of Portici, judged very properly, that to put an end to the
alarm this report had spread over the country, it was necessary to send up people he could trust, and on whose veracity
he might depend. Accordingly the next day, which was the
16th of July, Signor GUISEPPE SACCO went up, well attended,
and proved the gardener's assertion to be absolutely false, there
being only some little signs of mud from a deposition of the

rain water at the bottom of the crater. According to SACCO's account, which has been printed at Naples, the crater is of an irregular oval form, and, as he supposes (not having been able to measure it) of about a mile and an half in circumference; by my eye I should judge it to be more; the inside, as usual, in the shape of an inverted cone, the inner walls of which on the eastern side are perpendicular; but on the western side of the crater, which is much lower, the descent was practicable, and SACCO with some of his companions actually went down 176 palms, from which spot, having lowered a cord with a stone tied to it, they found the whole depth of the crater to be about 500 palms. But such observations on the crater of Vesuvius are of little consequence, as both its form and apparent depth are subject to great alterations from day to day. These curious observers certainly ran some risk at that time, since which such a quantity of scoriæ and ashes have been thrown up from the crater, and even so lately as the 15th of this month, as must have proved fatal to any one within their reach.

The 22d of July, one of the new craters, which is the nearest to the town of Torre del Greco, threw up both fire and smoke, which circumstance, added to that of the lava's retaining its heat much longer than usual, seems to indicate that there may still be some fermentation under that part of the volcano. The lava in cooling often cracks, and causes a loud explosion, just as the ice does in the Glaciers in Switzerland; such reports are frequently heard now at the Torre del Greco; and as some of the inhabitants told me, they often see a vapour issue from the body of the lava, and taking fire in air, fall like those meteors vulgarly called falling stars.

The darkness occasioned by the fall of the ashes in the Campagna Felice extended itself, and varied, according to the prevailing winds. On the 19th of June it was so dark at Caserta, which is 15 miles from Naples, as to oblige the inhabitants to light candles at mid-day; and one day during the eruption, the darkness spread over Eeneventum, which is 30 miles from Vesuvius.

The Archbishop of Taranto, in a letter to Naples, and dated from that city the 18th of June, said, " We are involved in a " thick cloud of minute volcanic ashes, and we imagine that ". there must be a great eruption either of Mount Etna, or of " Stromboli." The bishop did not dream of their having proceeded from Vesuvius, which is about 250 miles from Taranto. We have had accounts also of the fall of the ashes during the late eruption at the very extremity of the province of Lecce, which is still farther off; and we have been assured likewise, that those clouds were replete with electrical matter: at Martino, near Taranto, a house was struck and much damaged by the lightning from one of these clouds. In the accounts of the great eruption of Vesuvius in 1631, mention is made of the extensive progress of the ashes from Vesuvius, and of the damage done by the *ferilli*, or volcanic lightning, which attended them in their course.

I must here mention a very extraordinary circumstance indeed, that happened near Sienna in the Tuscan state, about 18 hours after the commencement of the late eruption of Vesuvius on the 15th of June, although that phænomenon may have no relation to the eruption ; and which was communicated to me in the following words by the Earl of Bristol, bishop of Derry, in a letter dated from Sienna, July 12th, 1794: " In

" the midst of a most violent thunder-storm, about a dozen
" stones of various weights and dimensions fell at the feet of
" different people, men, women, and children ; the stones are
" of a quality not found in any part of the Siennese territory;
" they fell about 18 hours after the enormous eruption of Ve-
" suvius, which circumstance leaves a choice of difficulties in
" the solution of this extraordinary phænomenon : either these
" stones have been generated in this igneous mass of clouds,
" which produced such unusual thunder, or, which is equally
" incredible, they were thrown from Vesuvius at a distance of
" at least 250 miles ; judge then of its parabola. The philoso-
" phers here incline to the first solution. I wish much, Sir, to
" know your sentiments. My first objection was to the fact
" itself ; but of this there are so many eye witnesses, it seems
" impossible to withstand their evidence, and now I am re-
" duced to a perfect scepticism." His lordship was pleased to
send me a piece of one of the largest stones, which when entire weighed upwards of five pounds ; I have seen another
that has been sent to Naples entire, and weighs about one
pound. The outside of every stone that has been found, and
has been ascertained to have fallen from the cloud near Sienna,
is evidently freshly vitrified, and is black, having every sign
of having passed through an extreme heat ; when broken, the
inside is of a light-grey colour mixed with black spots, and
some shining particles, which the learned here have decided
to be pyrites, and therefore it cannot be a lava, or they would
have been decomposed. Stones of the same nature, at least as
far as the eye can judge of them, are frequently found on
Mount Vesuvius ; and when I was on the mountain lately, I
searched for such stones near the new mouths, but as the soil

round them has been covered with a thick bed of fine ashes, whatever was thrown up during the force of the eruption lies buried under those ashes. Should we find similar stones with the same vitrified coat on them on Mount Vesuvius, as I told Lord BRISTOL in my answer to his letter, the question would be decided in favour of Vesuvius; unless it could be proved that there had been, about the time of the fall of these stones in the Sanese territory, some nearer opening of the earth, attended with an emission of volcanic matter, which might very well be, as the mountain of Radicofani, within 50 miles of Sienna, is certainly volcanic. I mentioned to his lordship another idea that struck me. As we have proofs during the late eruption of a quantity of ashes of Vesuvius having been carried to a greater distance than where the stones fell in the Sanese territory, might not the same ashes have been carried over the Sanese territory, and mixing with a stormy cloud, have been collected together just as hailstones are sometimes into lumps of ice, in which shape they fall; and might not the exterior vitrification of those lumps of accumulated and hardened volcanic matter have been occasioned by the action of the electric fluid on them? The celebrated Father AMBROGIO SOLDANI, professor of mathematics in the university of Sienna, is printing there his dissertation upon this extraordinary phænomenon; wherein, as I have been assured, he has decided that those stones were generated in the air independantly of volcanic assistance.

Until after the 7th of July, when the last cloud broke over Vesuvius, and formed a tremendous torrent of mud, which took its course across the great road between Torre del Greco and the Torre dell' Annunziata, and destroyed many vineyards,

the late eruption could not be said to have finished, although
the force of it was over the 22d of June, since which time the
crater has been usually visible. The power of attraction in
mountains is well known; but whether the attractive power
of a volcanic mountain be greater than that of any other
mountain, is a question : all I can say is, that during this last
eruption every watery cloud has been evidently attracted by
Vesuvius, and the sudden dissolution of those clouds has left
such marks of their destructive power on the face of the country all round the basis of the volcano as will not soon be erased.
Since the mouth of Vesuvius has been enlarged, I have seen a
great cloud passing over it, and which not only was attracted,
but was sucked in, and disappeared in a moment.

After every violent eruption of Mount Vesuvius, we read of
damage done by a mephitic vapour, which coming from under
the ancient lavas, insinuates itself into low places, such as the
cellars and wells of the houses situated at the foot of the volcano.
After the eruption of 1767, I remember that there were several
instances, as in this, of people going into their cellars at Portici,
and other parts of that neighbourhood, having been struck down
by this vapour, and who would have expired if they had not
been hastily removed. These occasional vapours, and which
are called here *mofete*, are of the same quality as that perma-
nent one in the Grotta del Cane, near the lake of Agnano, and
which has been proved to be chiefly fixed air. The vapours,
that in the volcanic language of this country are called *fumaroli*, are of another nature, and issue from spots all over the
fresh and hot lavas whilst they are cooling; they are sulphureous and suffocating, so much so that often the birds that are
flying over them are overpowered, and fall down dead; of

which we have had many examples during this eruption, particularly of wood pigeons, that have been found dead on the lava. These vapours deposite a crust of sulphur, or salts, particularly of sal ammoniac, on the scoriæ of the lava through which they pass; and the small crystals of which they are composed are often tinged with a deep or pale yellow, with a bright red like cinnabar, and sometimes with green, or an azure blue. Since the late eruption, many pieces of the scoriæ of the fresh lava have been found powdered with a lucid substance, exactly like the brightest steel or iron filings.

The first appearance of the *mofete*, after the late eruption, was on the 17th of June, when a peasant going with an ass to his vineyard, a little above the village of Resina, in a narrow hollow way, the ass dropped down, and seemed to be expiring ; the peasant was soon sensible of the mephitic vapour himself, and well knowing its fatal effects, dragged the animal out of its influence, and it soon recovered. From that time these vapours have greatly increased, and extended themselves. There are to this day many cellars and wells, all the way from Portici to Torre dell' Annunziata, greatly affected by them. This heavy vapour, when exposed to the open air, does not rise much more than a foot above the surface of the earth, but when it gets into a confined place, like a cellar or well, it rises and fills them as any other fluid would do ; having filled a well, it rises above it about a foot high, and then bending over, falls to the earth, on which it spreads, always preserving its usual level. Wherever this vapour issues, a wavering in the air is perceptible, like that which is produced by the burning of charcoal ; and when it issues from a fissure near any plants or vegetables, the leaves of those plants are seen to move, as if, they were agitated by a gentle wind. It is extraordinary, that although there does not appear to be any poisonous quality in this vapour, which in every respect resembles fixed air, it should prove so very fatal to the vineyards, some thousand acres of which have been destroyed by it since the late eruption; when it penetrates to the roots of the vines, it dries them up, and kills the plant. A peasant in the neighbourhood of Resina having suffered by the *mofete*, which destroyed his vineyards in the year 1767, and having observed then that the vapour followed the laws of all fluids; made a narrow deep ditch all round his vineyard, which communicated with ancient lavas, and also to a deep cavern under one of them; the consequence of his well reasoned operation has been, that although surrounded at present by these noxious vapours, and which lie constantly at the bottom of his ditch, they have never entered his vineyard, and his vines are now in a flourishing state, whilst those of his neighbours are perishing. Upwards of thirteen hundred hares, and many pheasants and partridges, overtaken by this vapour, have been found dead within his Sicilian Majesty's reserved chases in the neighbourhood of Vesuvius; and also many domestic cats, who in their pursuit after this game fell victims to the *mofete*. A few days ago a shoal of fish, of several hundred weight, having been observed by some fishermen at Resina in great agitation on the surface of the sea, near some rocks of an ancient lava that had run into the sea, they surrounded them with their nets, and took them all with ease, and afterwards discovered that they had been stunned by the mephitic vapour, which at that time issued forcibly from underneath the ancient lava into the sea. I have been assured by many fishermen, that during the force of the late eruption the fish had totally abandoned the coast from Portici to the Torre dell' Annunziata, and that they could not take one in their nets nearer the shore than two miles. The divers there, who fish for the *ancini* (which we call sea eggs) and other shell fish, likewise told me, that for the space of a mile from that shore, since the eruption, they have found all the fish dead in their shells, as they suppose either from the heat of the sand at the bottom of the sea, or from poisonous vapours. The divers at Naples complain of their finding also many of these shell fish, or as they are called here in general terms, *frutti di mare*, dead in their shells.

I thought that these little well attested facts might contribute to show the great force of the wonderful chemical operation of nature that has lately been exhibited here. The *mofete*, or fixed air vapours, must certainly have been generated by the action of the vitriolic acid upon the calcareous earth, as both abound in Vesuvius. The sublimations, which are visibly operating by the chemistry of nature all along the course of the last lava that ran from Vesuvius, and particularly in and about the new mouths that have been formed by the late eruption on the flanks of the volcano, having been analyzed by Signor DOMENICO TOMASO, an ingenious chemist of Naples, and whose experiments, and the result of them, are now published, have been found to be chiefly sal ammoniac, mixed with a small quantity of the calx of iron : but not to betray my ignorance on this subject, and pretending to nothing more than the being an exact ocular observer, I refer you to the work itself, which accompanies this letter. Many hundred weight of the Vesuvian sal ammoniac have been collected on the mountain since the late eruption by the peasants, and sold at

Naples to the refiners of metals ; at first it was sold for about six pence a pound, but, from its abundance, the price is now reduced to half that money; and a much greater quantity must have escaped in the air by evaporation.

The situation of Mount Vesuvius so near a great capital, and the facility of approaching it, has certainly afforded more opportunities of watching the operations of an active volcano, and of making observations upon it, than any other volcano on the face of the earth has allowed of. The Vesuvian diary, which by my care has now been kept with great exactness, and without interruption for more than 15 years, by the worthy and ingenious Padre ANTONIO PIAGGI, as mentioned in the beginning of this letter, and which it is my intention to deposite in the library of the Royal Society, will also throw a great light upon this curious subject. But as there is every reason to believe, with SENECA,* that the seat of the fire that causes these eruptions of volcanoes is by no means superficial, but lies deep in the bowels of the earth, and where no eye can penetrate, it will, I fear, be ever much beyond the reach of the limited human understanding to account for them with any degree of accuracy. There are modern philosophers who propose, with as great confidence, the erecting of conductors to prevent the bad effects of earthquakes and volcanoes, and who promise themselves the same success as that which has attended Doctor FRANKLIN's conductors of lightning ; for, as they say, all proceed from one and the same cause, *electricity*. When we reflect how many parts of the earth already inhabited have evidently been thrown up from the bottom of the

---

* " Non ipse ex se est, sed in aliqua inferna valle conceptus exæstuat, et alibi pas-
" citur ; *in ipso monte non alimentum habet, sed viam.*"—SENECA, Epist. 79.

sea by volcanic explosions, and the probability of there being a much greater portion under the same predicament, as yet unexplored, the vain pretensions of weak mortals to counteract such great operations, carried on surely for the wisest purposes by the beneficent Author of nature, appear to me to be quite ridiculous.

Let us then content ourselves with seeing, as well as we can, what we are permitted to see, and reason upon it to the best of our limited understandings, well assured that whatever is, is right.

The late sufferers at Torre del Greco, although his Sicilian Majesty, with his usual clemency, offered them a more secure spot to rebuild their town on, are obstinately employed in rebuilding it on the late and still smoking lava that covers their former habitations ; and there does not appear to be any situation more exposed to the numerous dangers that must attend the neighbourhood of an active volcano than that of Torre del Greco. It was totally destroyed in 1631; and in the year 1737 a dreadful lava ran within a few yards of one of the gates of the town, and now over the middle of it ; nevertheless, such is the attachment of the inhabitants to their native spot, although attended with such imminent danger, that of 18000 not one gave his vote to abandon it. When I was in Calabria, during the earthquakes in 1783, I observed in the Calabrese the same attachment to native soil ; some of the towns that were totally destroyed by the earthquakes, and which had been ill situated in every respect, and in a bad air, were to be rebuilt ; and yet it required the authority of government to oblige the inhabitants of those ruined towns to change their situation for a much better.

Upon the whole, having read every account of the former eruptions of Mount Vesuvius, I am well convinced that this eruption was by far the most violent that has been recorded after the two great eruptions of 79 and 1631, which were undoubtedly still more violent and destructive. The same phænomena attended the last eruption as the two former above mentioned, but on a less scale, and without the circumstance of the sea having retired from the coast. I remarked more than once, whilst I was in my boat, an unusual motion in the sea during the late eruption. On the 18th of June I observed, and so did my boatman, that although it was a perfect calm, the waves suddenly rose and dashed against the shore, causing a white foam, but which subsided in a few minutes. On the 15th, the night of the great eruption, the corks that support the nets of the royal tunny fishery at Portici, and which usually float upon the surface of the sea, were suddenly drawn under water, and remained so for a short space of time, which indicates, that either there must have been at that time a swell in the sea, or a depression or sinking of the earth under it.

From what we have seen lately here, and from what we read of former eruptions of Vesuvius, and of other active volcanoes, their neighbourhood must always be attended with danger; with this consideration, the very numerous population at the foot of Vesuvius is remarkable. From Naples to Castel-a-mare, about 15 miles, is so thickly spread with houses as to be nearly one continued street, and on the Somma side of the volcano, the towns and villages are scarcely a mile from one another; so that for thirty miles, which is the extent of the basis of Mount Vesuvius and Somma, the population may be perhaps more numerous than that of any spot of a like extent in Europe, in spite of the variety of dangers attending such a situation.

With the help of the drawings that accompany this account of the late eruption of Vesuvius, and which I can assure you to be faithful representations of what we have seen, I flatter myself I shall have enabled you to have a clear idea of it; and I flatter myself also, that the communication of such a variety of well attested phænomena as have attended this formidable eruption, may not only prove acceptable, but useful to the curious in natural history.

I have the honour to be, &c.

WM. HAMILTON.

IN a subsequent letter from Sir WILLIAM HAMILTON to Sir JOSEPH BANKS, dated *Castel-a-mare*, anciently *Stabiæ*, Sept. 2, 1794, are the two following remarks to be added to this paper.

1. Within a mile of this place the *mofete* are still very active, and particularly under the spot where the ancient town of Stabiæ was situated. The 24th of August, a young lad by accident falling into a well there that was dry, but full of the mephitic vapour, was immediately suffocated; there were no signs of any hurt from the fall, as the well was shallow. This circumstance called to my mind the death of the elder PLINY, who most probably lost his life by the same sort of mephitic vapours, on this very spot, and which are active after great eruptions of Vesuvius.

2. Mr. JAMES, a British merchant, who now lives in this neighbourhood, assured me that on Tuesday night, the 17th of

June, which was the third day of the eruption of Mount Vesuvius, he was in a boat with a sail, near Torre del Greco, when
the minute ashes, so often mentioned in my letter, fell thick ;
and that in the dark they emitted a pale light like phosphorus,
so that his hat, those of the boatmen, and the part of the sails
that were covered with the ashes, were luminous. Others
have mentioned to me the having seen a phosphoric light on
Vesuvius after this eruption ; but until it was confirmed to
me by Mr. JAMES, I did not choose to say any thing about it.

EXPLANATION OF THE PLATES.

Tab. V. Is a view of the eruption of Mount Vesuvius on
the night of the 15th of June, 1794, taken from S. Lucia at
Naples, when the eruption was in its greatest force.

Tab. VI. Is a view of the lava that destroyed the town of
Torre del Greco, taken from a boat on the sea near that town,
about five o'clock in the morning of the 16th of June, and
whilst the lava was still advancing in the sea. The rocks, on
which are two figures near the boat, were formed by a lava
that ran into the sea during a former eruption of Mount Vesuvius.

Tab. VII. Is a view of the enormous cloud of smoke and
ashes, replete with *ferilli,* or volcanic lightning, which first
threatened destruction to the town of Naples on the 18th of
June ; and afterwards, from the impulse of the sea wind, bent
over the mountain of Somma, and poured its destructive con-
tents on the towns situated at the foot of that mountain, beat-
ing in the roofs of the houses, and involving all the inhabitants of the Campagna Felice in darkness and danger. This

view was taken from Naples, and gives a very good idea of the appearance of Mount Vesuvius, like a mole-hill, in comparison of the enormous mass that hung over it.

Tab. VIII. Is a view of Mount Vesuvius, and of Somma, taken from Posilipo July 6th, 1794, when it could be clearly distinguished ; the dotted lines shew the form of the top of Vesuvius as it was before this eruption, and when the crater was only from A to B ; the present wide extended crater is sufficiently plain in the drawing not to need any further explanation ; the spot from whence the lava first issued the night of the 15th of June, is marked C.

These four very exact drawings were taken from nature by Signor XAVERIO GATTA, successor to Signor PIETRO FABRIS.

Tab. IX. Is a drawing made by the Padre ANTONIO PIAGGI at Resina, during the force of the eruption of the 15th at night ; and being within a mile and a half of the mountain, shews many particulars that escaped us, so much farther off at Naples ; but he was interrupted by the imminent danger of his situation, and his drawing is incomplete : it was with difficulty that his friends carried him off alive, being upwards of 80 years old, in the midst of a shower of heavy cinders and sulphureous ashes, an hour after the beginning of the eruption ; nor was he able to return to his house for many days. Nothing is necessary to be added to his Latin references to the drawing, but that Turris VIII. is *Torre del Greco*, and Retina, now *Resina*.

A. Montis vertex innubis, compositusque.

B. ad H. Sulci rudes inhianti terræ frequenter inscripti.

D. Ignei rivi fluentes Retinam versus.

E. Nitidissima flamma in cupressus formam altitudinem montis exsuperans.

F. Saxorum tempestas in altum a voraginibus erumpentium.

G. Lenis clivus igneum flumen in Retinam minantem avertens.

H. Semita ignei torrentis incredibili rapiditate Turrim VIII. invasuri.

I. Arbores, et vineta simplici illius afflatu a longe micantia.

K. Turris VIII. quæ Herculanio successisse creditur.

L. S'ᵗ. Mariæ Apulianæ templum.

M. Retina templo adhærens, recenter constructa, ab illo usque ad mare.

N. Porticus : nova item constructio Neapolim versus, unum corpus cum Retina efficiens.

O. Leucopetra.

P. Massa.

Q. Trochlea.

R. Sᵈⁱ. Sebastiani vicus.

S. Fumus lapillis, asperis arenis, et aqua marina confertus in pluviam solutus.

Tab. X. Plan of the city of Torre del Greco, destroyed in great part by the lava which ran in the night of the 15th of June, 1794.

Tab. XI. Map of Mount Vesuvius and the adjacent places, with the course of the lava.

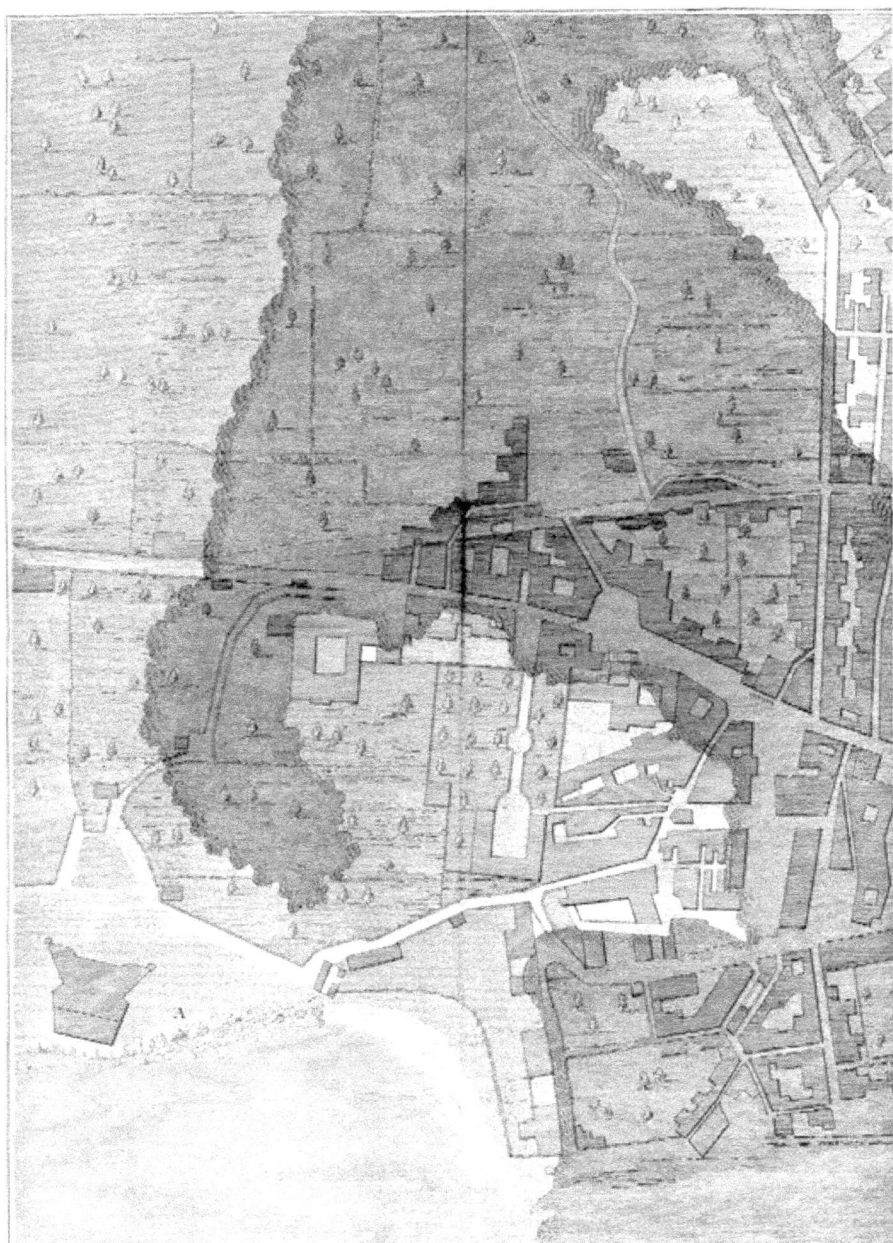

A

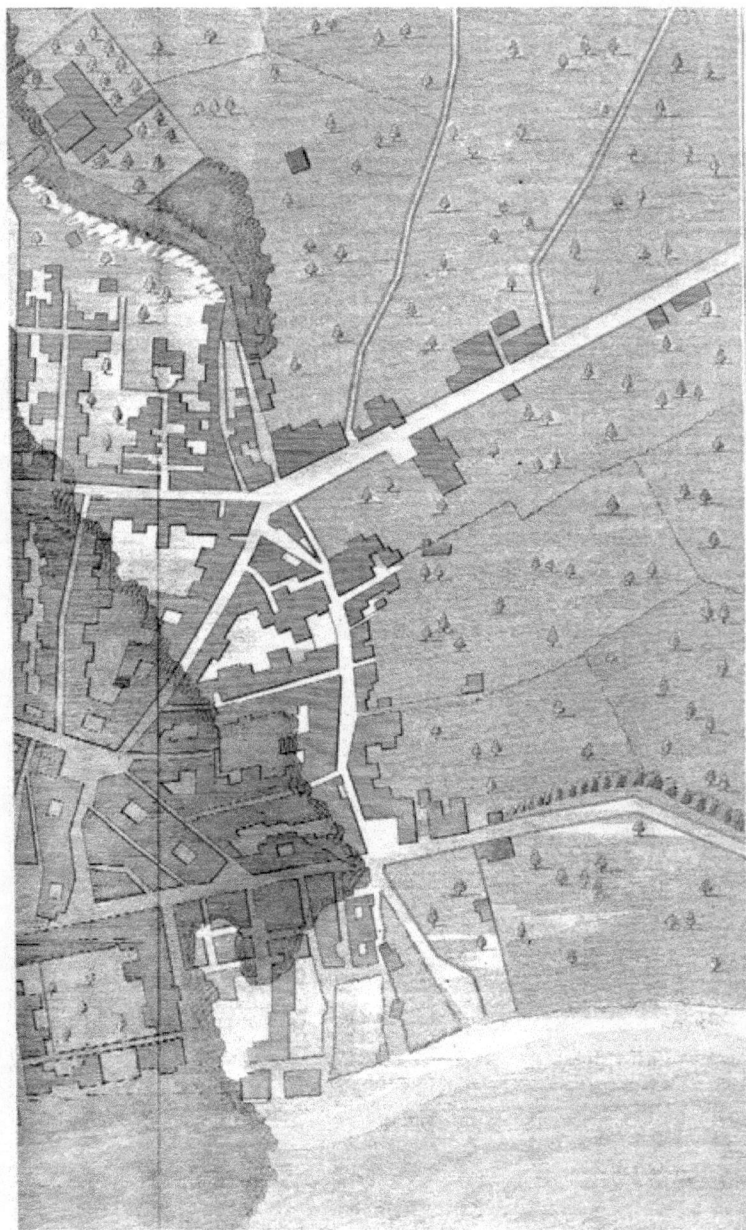

The parts shaded dark are those covered by the late Lava.— N.

Scale of 7000 Neapolitan Palms

A. Shews where the Lava of 1631 entered the Sea.

A.B. The late Current of Lava, with its branches, shaded darker than th...

C.C.C.C. Course of the Torrent of 20.th and 21.st of June. D.D.D. Torrent of the 5.th o...

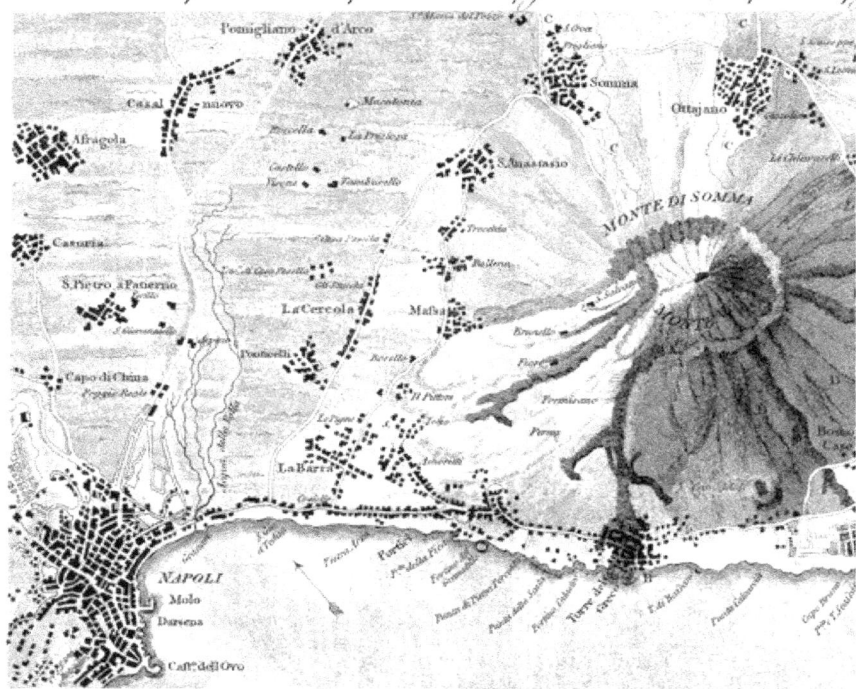

S.ta Maria del Pozzo

Pomigliano d'Arco

Somma

Ottajano

Casal nuovo

Macalonia

Afragola

Piccola La Preziosa

S. Anastasio

MONTE DI SOMMA

Castello Vierra

Tombarello

Casoria

Trecchia

S. Pietro a Paterno

Col. d. Caza Puella

Gli Zazoli

Ballora

La Cercola

Massa

Ponticelli

Rocello

Capo di Chiua
Poggio Reale

Il Pileri

Le Pigne

Toke

NAPOLI

La Barra

Porici

Molo

Darsena

Torre d...

Cast.o dell Ovo

Italian   1   2   3   4   5   6

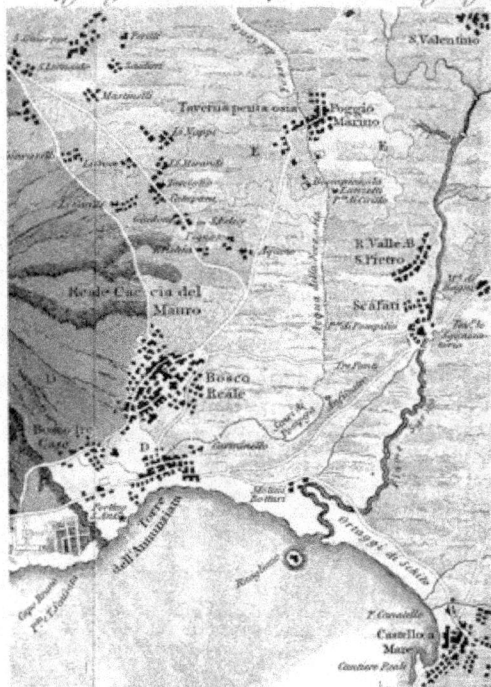

... ran the course of former Currents.
...e 5.th of July. E.E. Torrent of the 5.th 6.th & 7.th of July.

6   8 Miles.